I0605676

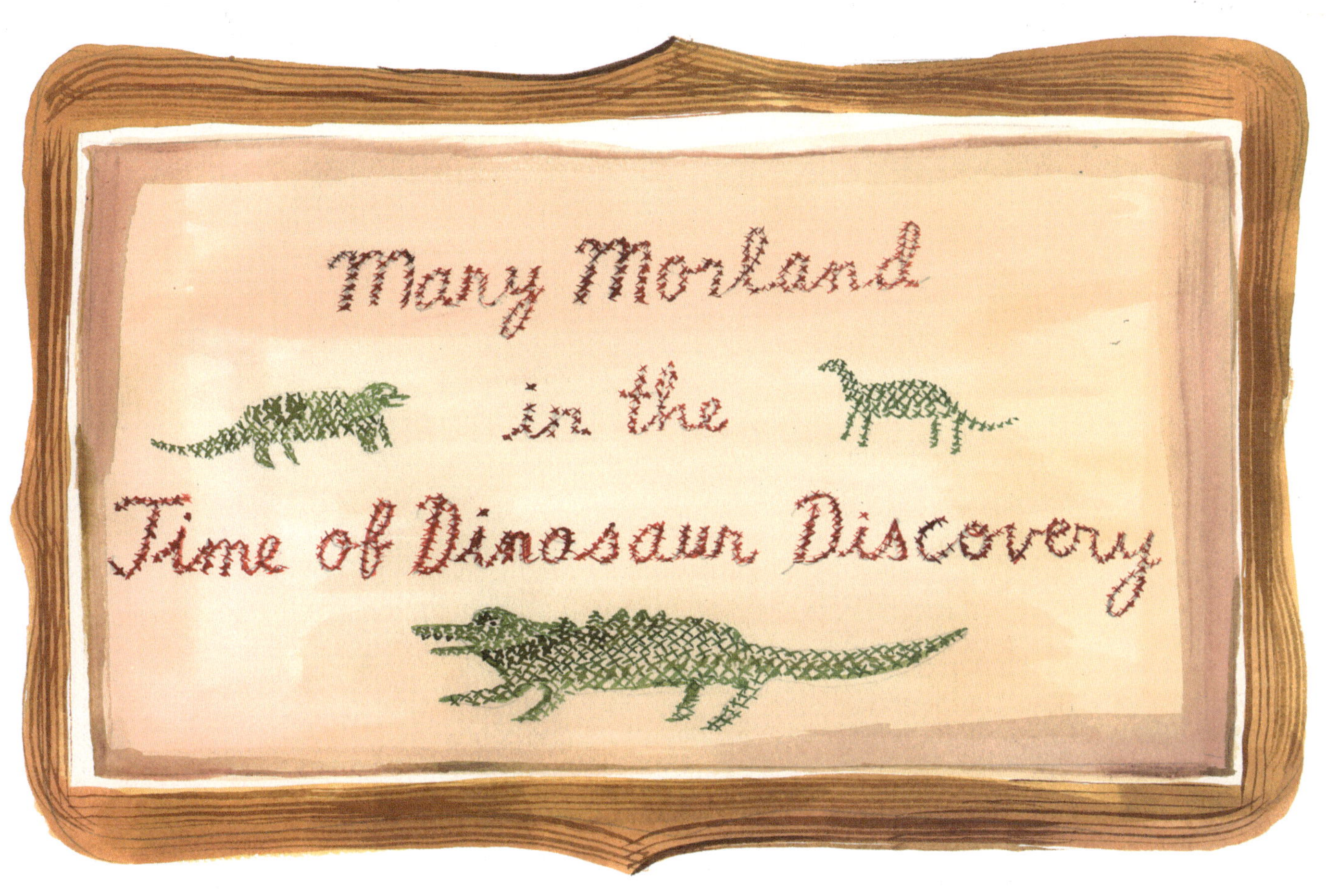
Mary Morland
in the
Time of Dinosaur Discovery

Mary Morland *in the Time of* Dinosaur Discovery

Written by
Jane Kurtz

Illustrated by
Giselle Potter

Beach Lane Books
New York Amsterdam/Antwerp London Toronto Sydney/Melbourne New Delhi

Mary Morland was born in a time of ribbons and lace, when small girls were seen as dainty, delicate decorations to make a room pretty.

Did she perch
on her chair
like a prim little miss?

Well, what would you do if the whole world was waiting outside?

Mary tromped around, exploring.

Wouldn't you?

When Mary's mother died, her father sent her to live with friends who had no children. Did Mary feel sad about that?

Probably.

Wouldn't you?

Professor Pegge taught her to look closely at plants and animals, rocks and fossils.

And did she?

Yes!

Her journals captured everything that fascinated her.

When Mary was a teenager,
did she giggle and say,
"What a cork-brain I am"?

Did she spend her days
thinking about
good manners and . . .

HATS?

NOPE.

Whenever she could, she hopped into her carriage, clicked her tongue to her white donkey, and went off to collect shells and fossils.

She also wrote letters to Georges Cuvier, a famous French scientist who had studied fossilized bones in Paris and insisted that ancient reptiles once swam in the sea and flew through the air.

Astonishing!

Mary sent specimens she found and drawings she made to Georges Cuvier. Was he impressed?

Yes.

So she sent him more.

A few years later, everyone said the Number One Thing for a young lady was to find a good husband.

Did Mary spend all her time whirling around ballrooms, smiling shyly at young men?

Did she collect slices of bridecake to put under her pillow so she would dream of her future sweetheart?

She collected rare types of sea sponges.
She made models of fossils, labeled fossils,
and mended broken fossils.

She treasured Professor Pegge's collection of minerals, which became hers after he died.

One day Mary was riding in a carriage and reading a thick book by Georges Cuvier. Across from her sat a man with the same thick book. "You must be Miss Morland!" he said.

The man's name was William Buckland. Once he had been Professor Pegge's student. Now he was England's best fossilist.

When the famous French scientist Georges Cuvier had visited England, it was William Buckland who got to show him the collections of English fossils Buckland was studying. "Ah!" Cuvier had said.

The bones must have come from a giant *land* reptile! Buckland chose a name that meant "great lizard":

Megalosaurus.

Did William Buckland rush out and share *Megalosaurus* with the world?

Nope.

(No one is quite sure why not.)

During their bumpy carriage ride, Mary Morland learned that William Buckland had gotten a scientific medal when he studied a cave full of white blobs and old bones belonging to animals no one had ever seen in England. Could hyenas have pulled the bones into the cave? Time for an experiment! He brought home a hyena named Billy, and YES! Billy's poo matched the white blobs in the cave.

Did Mary shriek and say,
"Sir! What an indelicate conversation!"?

It seems she did not. In fact, she agreed to make drawings of hyena jaws for a book he wanted to write.

Of course, William Buckland then wanted her to draw something else:

Megalosaurus.

When she saw the fossilized *Megalosaurus* bones, did she say, "I wonder if they might be from a dinosaur"?

No, she did not.

The word "dinosaur" hadn't been invented yet.

She *did* agree to draw his collection of *Megalosaurus* fossils.

When Mary finished, William Buckland finally decided it was time to introduce *Megalosaurus* to the world—and he wanted to do it with a brilliant flourish.

One night, when all of England's geologists were gathered in a room, he gave a lecture describing this huge ancient land-crawling reptile.

It was an astonishing idea for men who knew about giant ancient *sea* creatures but not giant ancient *land* creatures.

William Buckland even had illustrations of the fossils to pass around!

Of course, the illustrations had been made by Mary.

Megalosaurus became the first dinosaur to have a name—even before the word "dinosaur" was invented.

Next thing we know, William Buckland and Mary Morland were married and going on a honeymoon trip around Europe to meet famous scientists and visit important spots where fossils had been found.

Then did Mary settle down to needlework and the cutting of roses? Did she turn her back on all her drawings and collections forever?

What would you do?

Well, what WOULD you do?

Mary chose to be in charge of a household of chaos. Every room was stuffed with science books and papers and dusty rocks. One time, she said, she tried to clean things up.

Never again.

If you tiptoed down the hall, past cages full of snakes and green frogs, would you see an elegant and neat dining room?

Not at all.

Is that a pony running around the table with three of Mary's laughing children on its back?

Yes. Yes, it is.

Mary and William experimented together.

They sometimes stayed up all night while he talked and she wrote things down. She helped turn his writing into books that also included her drawings.

When his scientific friends needed to see something from his collections, they turned to her.

Did William say, "I am so fortunate that my wife is not a cork-brain"?
Did he say, "Let us make sure your name is on these papers
so the world will know your part"?
Did he say, "Without you, my dear, my work would be a shemozzle"?

No.
No, he did not.

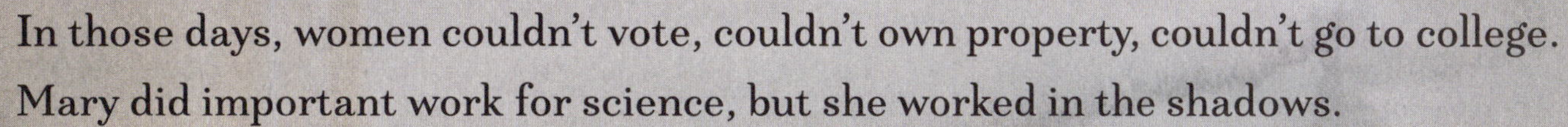

In those days, women couldn't vote, couldn't own property, couldn't go to college. Mary did important work for science, but she worked in the shadows.

What did Mary think of that?
She did not write down her reaction.
Neither did anybody else.

But she kept learning and passing her enthusiasms on.

She painted illustrations of living animals to help William with his teaching.

She made paper globes for the local school.

She taught her own children—and her son Frank later became a celebrity who passed on his love of science to thousands of other people.

Until the very day she died, Mary Morland found the world a fascinating place and worked hard to learn more about it.

Wouldn't you?

Author's Note

Mary Morland (1797–1857) was not the only woman to be part of the early story of dinosaurs. You might know about Mary Anning, who discovered the first *Plesiosaurus*. Georges Cuvier thought Mary Anning's skeleton might have been a fake. In William Buckland's words, it looked like something with a lizard's head, a crocodile's teeth, a chameleon's ribs, a whale's paddles, and the neck of a serpent. William Buckland went to see the *Plesiosaurus* skeleton and help a friend buy it and take it back to London. Then Mary Morland made a drawing that convinced Georges Cuvier to take the fossil seriously.

For the most part, this period of putting together dinosaur clues belonged to men. Women weren't allowed to be Fellows in the Geological Society of London. They had to experience the life of science by educating themselves and with the help of male friends and husbands.

One grand night (with no women present), the Geological Society of London got to see Mary Anning's *Plesiosaurus* for the first time. That night William Buckland also read a paper introducing the world to the dinosaur now called *Megalosaurus bucklandii* and showed beautifully detailed drawings by Mary Morland. A few years later, when William Buckland came up with a name for fossilized dung—coprolite—he gave Mary Anning credit for discovering coprolites in the cliffs near her home in Lyme Regis and figuring out what they must be. The second dinosaur to get a name was *Iguanodon*, and the man who named it, Gideon Mantell, credited his wife, Mary Ann, for being part of the discovery. Later the scientist Richard Owen put those two fossils (along with a third) into a group: Dinosauria, or dinosaurs.

It took courage for these early fossilists to trust what they were seeing and draw conclusions that guided all the dinosaur discoveries that happened next in the world. Georges Cuvier, for example, came up with this helpful theory: Skulls always attached to backbones. Legs always attached to hips. He became skilled at drawing conclusions from bits of bones and teeth. These scientists were exploring in a time when people were used to having the church—not their observations—tell them what was true.

Mary would stay up with William Buckland when everyone else went to sleep, discussing his ideas and taking dictation. Their son Frank became a scientist who wrote with lively humor about all kinds of animals around the world. Frank and his sister Elizabeth are the main reasons we know details of their mother's life and work. Women didn't get to have the spotlight in those days. But at least three English women—Mary Anning, Mary Ann Mantell, and Mary Morland Buckland—all had a part in the dinosaur discovery story.

Selected Sources

Buckland, Frank. *Curious Men.* McSweeney's Books, 2008.

Chapman, Allan. *Caves, Coprolites and Catastrophes: The Story of Pioneering Geologist and Fossil-Hunter William Buckland.* SPCK Publishing, 2020.

Dolnick, Edward. *Dinosaurs at the Dinner Party: How an Eccentric Group of Victorians Discovered Prehistoric Creatures and Accidentally Upended the World.* Scribner, 2024.

Emling, Shelley. *The Fossil Hunter: Dinosaurs, Evolution, and the Woman Whose Discoveries Changed the World.* Saint Martin's Griffin, 2009.

Gordon, Elizabeth Oke. *The Life and Correspondence of William Buckland.* Murray, 1894.

University of Oxford Museum of Natural History. "Mary Morland." Accessed April 14, 2025. https://oumnh.ox.ac.uk/mary-morland-online-exhibition.

More than a Dodo: Oxford University Museum of Natural History Blog. "Of Jumping Mice and Megalosaurus." Posted April 4, 2024. https://morethanadodo.com/2024/04/04/of-jumping-mice-and-megalosaurus/.

Vickery, Amanda. *The Gentleman's Daughter: Women's Lives in Georgian England.* Yale University Press, 1998.

Resources for Curious Young Readers

Blackford, Cheryl. *Fossil Hunter: How Mary Anning Changed the Science of Prehistoric Life.* Clarion Books, 2022.

Kerley, Barbara. *The Dinosaurs of Waterhouse Hawkins.* Scholastic Press, 2001.

Lendler, Ian. *The First Dinosaur: How Science Solved the Greatest Mystery on Earth.* Margaret K. McElderry Books, 2019.

For all the determined, smart women in my life,
including my writing buddies, my mom, my sisters,
my daughter, and my granddaughter
—J. K.

For Pia and Izzy,
who also love collecting shells
and stones to paint
—G. P.

BEACH LANE BOOKS • An imprint of Simon & Schuster Children's Publishing Division • 1230 Avenue of the Americas, New York, New York 10020 • For more than 100 years, Simon & Schuster has championed authors and the stories they create. By respecting the copyright of an author's intellectual property, you enable Simon & Schuster and the author to continue publishing exceptional books for years to come. We thank you for supporting the author's copyright by purchasing an authorized edition of this book. • Book design by Lauren Rille • For information about special discounts for bulk purchases, please contact Simon & Schuster Special Sales at 1-866-506-1949 or business@simonandschuster.com. • Simon & Schuster strongly believes in freedom of expression and stands against censorship in all its forms. For more information, visit BooksBelong.com. • The Simon & Schuster Speakers Bureau can bring authors to your live event. For more information or to book an event, contact the Simon & Schuster Speakers Bureau at 1-866-248-3049 or visit our website at www.simonspeakers.com. • The text for this book was set in Marcia. • The illustrations for this book were rendered in watercolor. • Manufactured in China • 1025 SCP • First Edition • 10 9 8 7 6 5 4 3 2 1 • CIP data for this book is available from the Library of Congress. • ISBN 9781665955546 • ISBN 9781665955553 (ebook)